More Praise for *Between River & Street*

"With *Between River & Street*, Scott T. Starbuck is deep listener, ace observer, shape-shifting storyteller. He's historian, philosopher, climatologist, ecologist, lover, rememberer. He's salmon, and salmon fisherman. Those who know Northwest rivers will want to pass this passionate book to friends. Those streetwise to our environmentally-challenged world will catch these smart, tough poems in order to release as seems right. Like the late Oregon poet, William Stafford, Starbuck writes of place with integrity, authenticity, humanity."

—Ken Waldman, author of *The Writing Party* and *Leftovers and Gravy*

"Scott T. Starbuck's poems are Oregon poems—humble yet heartfelt, all the way to the bone. Like filets of rosy salmon flesh, cooked on bonfire coals; these words lift easily from the carcass of felt sense. *Between River & Street* is a collection of mostly short but never simple poems, reflecting a life lived with daily presence and purpose. Starbuck's poems attend to fish and to fish stream—the moss and fern, salal and spruce, wet stone and rivulet—but also the hominids, on the shores and in the quik marts and cafés of classic hometowns from Astoria south to the Siskiyou. These pages honor Oregon existence: good folk whose unfettered reverence plays potent role in the annual cycle of lives lived out in place. Readers, be ready; pull on your waders, step into the stream. Starbuck casts words like caddis flies; these poems hit the heart, like a trout hits the hook, lies briefly in careful hand, caught—and released into gratitude and requisite grief. Travel down the coast and up tributaries of the inland soul. These are poems of a citizen who is denizen, engaged in each damp day, and witness to those moments when clouds part and sun shimmers with 'shadow on the once-magic-waters.'"

—Nancy Cook, author of *Siltwater*,
a collection of essays, former editor of *RAIN Magazine*

"If you're not concerned about the imminent collapse of the ecosystems of earth, then you should be. Such is the overarching theme of Scott T. Starbuck's *Between River & Street*. Awareness is a call to action, to give up our citified car culture and rediscover our original home. About that home, Scott celebrates the natural world of fish (especially fish), birds, animals, and the manifold flora of his local ground, the Columbia Gorge and Pacific Coast. Scott's poems tell of some people unaware of that world, while other encounters yield insight on how we really live on earth. Scott's brief poems carry hope for the human future."

—Bill Siverly, author of *Nightfall*,
co-editor of *Windfall: A Journal of Poetry of Place*

Between River & Street

Between River & Street

Scott T. Starbuck

MoonPath Press

Poetry
ISBN 978-1-936657-58-2

Cover photo *Grayscale photo of bridge under cloudy sky*
by Joseph Miguel
courtesy of scop.io

Author photo by Suzette Starbuck

Book design by Tonya Namura, using GriffosFont
and Gentium Basic.

MoonPath Press, an imprint of Concrete Wolf Poetry Series,
is dedicated to publishing the finest poets
living in the U.S. Pacific Northwest.

MoonPath Press
PO Box 445
Tillamook, OR 97141

MoonPathPress@gmail.com

http://MoonPathPress.com

Acknowledgments

Grateful acknowledgment is made to the following in which these poems first appeared, or are forthcoming, sometimes in earlier versions.

Blue Begonia Press Poetry Pole in Selah, Washington: "Dream of Customer Service"

Cascadia Rising Review: "River Reflections"

Cirque: "Man and Black Lab by Fire"

North Coast Squid: "Deep Listening on Oregon Coast" and "*Salal* 2019 Event in Rose Center at Lower Columbia College"

The Oregonian: "Dream of Peasants' Revolt," "The Hunger," "Remembering Rocky Creek" (published under the title "Once, a Red Tide"), "Whale Eggs"

The Portland Review: "Hat Haiku"

Rain Magazine: "At Astoria Bistro," "December Sun," "Pigeons Can't Fly," " Oregon Traitor"

Salal Review: "What the Salmon Said"

Stringtown: "Fishing at Age 50"

Verse Wisconsin: "Maynard's Love"

"The Hunger" and "A Spell of Birds and Fish" appeared in the chapbook *The Eyes of Those Who Broke Free* by Pudding House Publications.

"Memory," "Mushroom Hunting on Black Friday," and "Oregon City Flood Year 1964" (previously titled "Flood Year") appeared in *Carbonfish Blues.*

"Remembering Rocky Creek" appeared in *Lost Salmon* under the title "At Rocky Creek."

Table of Contents

I. What the Salmon Said

II. Remembering Rocky Creek

III. In Praise of Wild Rivers

Between River & Street

I. What the Salmon Said

November

I’m ready for unmarked snow,
water-ouzel silence.

I’m wading for clues
of first steelhead,

blood on rock,
gills or guts in shallows.

I’m meeting Derrick the logger,
Hal the heron,

water reflections of lost friends
and caught fish,

any blind spot waiting to show
what was, is, will be

and can never be,
here to honor original fire

returning fish
and everything real.

Flamekeepers

weren't men but ravens
as in Tlingit legends,

salmon, halibut, ling cod,
black cod, sea bass, spot prawns,

muskeg, Sitka spruce,
Western hemlock, bears,

last refuges of hidden
and visible sacred.

Then

After days of trolling
with no strikes
there was a moment
when a giant slab of silver
chinooked my hook
and was gone.

Strawberry Pickers, Oregon, 1970

Jerry Hayes and I smuggled line,
hooks, worms, in Band-Aid cans
to catch tiny rainbows
in nearby creek,

with no way to take them home
and no other way
to get fish.

River Reflections

When I was 18 a famous writer called me
after I entered a story contest in the *Oregonian's*
Northwest Magazine.

"You didn't win," he started,
"but I saw your piece in *Salmon Trout Steelheader*
and have a friend from out of town."

"Can you give us a few fishing spots?"
I gave him a few spots and tips
reflecting how even the most accomplished

turn to locals in the end
as it will be with climate change,
formerly secluded mansion millionaire

hungry beyond belief
asking some rough or urchin
where he can catch a fish.

"Who Owns the Salmon?"

asks my nephew, Brandon,
and I say he does
since he just bought a tag.

Or, you could say the river does.
Or, you could say they own themselves,
same as you.
Or, you could say watery Earth.
Or, you could say glittery Milky Way Galaxy.
Or, you could say that which gives rise to glittery
galaxies.
Or, you could say that which gives rise to
that which gives rise to glittery galaxies.

Then again, you could also say
that which gives rise to
that which gives rise to
that which gives rise to glittery galaxies.

By the time I get to the Creator
Brandon is asleep
or pretends to be.

Gramps

Forty fathoms off Otter Crest,
charter trolling for coho salmon,
the man they refer to as Gramps
wants a bobber and worm
like for Kansas catfishing.

"Ain't got no bobber
or worm," I report.
"Ha. And you call yourself
a fisherman?" he snarls.
"Fish on!" someone screams

and, of course, it's Gramps' rod.
He sets the hook so hard
it will either be a broken line
or missed fish.
"Got away," he says. "Dull hook."

"Can't set the hook
on these salmon," I suggest,
"the boat's moving too fast."
"Don't tell me
how to fish, Sonny," he says

"I've caught Kansas catfish so big
they make Oregon salmon
look like bait."

Deep Listening on Oregon Coast

A Shell Station attendant offers advice
about surf perch fishing
in "patience" and "time"

—word-medicine
for an unspoken thorn
that has nothing to do with fishing.

What the Salmon Said

I wander as far as Japan
but come home
when time is right in fiery maples
to birthgravel in finned song
of ancestors.

There is joy in spawning,
dying, completing circle
ordained by wise maker
of all real circles.

Oregon Traitor

Jake refused to fish his father's spinners
because he caught more on another brand.

"Reality doesn't care about family," he said,
"and neither do these salmon."

"Does it care about you borrowing my truck?"
asked his dad, staring him down.

Father and son cursed each other in tidewater
all spring under stoic herons,

soaring eagles.

Crow

is upset he's not eagle
so complains to Birdmaker
and Birdmaker says
"You think sky is about you
and it is, but
it is also about everyone
and everything
so if I didn't make you crow
and eagle eagle
there could be no bird,
Birdmaker, or even sky.

"Crow, you are as beautiful
in my eyes as eagle."

The Hunger

You led me to night hawk
circling orchards for mice,

lantern moon,

diamond eyes
of 80-year-old sturgeon
cruising river bank
like forgotten gods.

We walked in the hour
skeleton oaks
reached out of hiding
to touch,

just a short time
before tide moved in
and this place became
invisible.

Sixth Sense

Standing on Chicken Point with other sea captains,
watching waves build, heave, crash,
I am first to cancel my Depoe Bay charter
as I have no wife, kids, mortgage, boat payment
gluing my eyes to North Reef, South Reef,
and rift between money and scratched hull or worse.
I am deep listening for ancient sixth sense,
maybe whisper of angel, siren, or
that mysterious self we all have who
most of the time knows much more than
any of us admit.

Inventing a New Language

requires deep listening
and long walks by still
or moving water.

It requires wisdom
of stomach, feet, nose,
tongue, ears and intuition.

It requires unlearning
lesser languages
that functioned when

you left your child-self
behind in autumn leaves
to navigate steel and stone

of men caught
in an exchange
they believed was real.

Post Office Bob

and I met at fish hole by accident.
"You go ahead," he said,

hiking back in Oregon rain
to white Ford Ranger.

"Wanna see steelhead?" he asked
later, then drove me

to seven finning a tailout
behind a log.

No Amway. No preaching.
Just gift.

Later, driving home
I recalled my frequent greed,

anger, impatience fishing
with strangers.

Abundant razor clams,
chanterelles, wild raspberries.

Ancient Sitka Spruce
watching.

Passing Over Los Angeles River on Amtrak #782

I saw a tire
round like Earth
but hollow too
like men.

This tire used to help
families move
from home
to school to work.

Now, it's the giant eye
of a dead beast
or portal
to concrete slab.

Long ago
Pacific salmon
were replaced here
by carp and trash.

Tongva and Chumash
by movie stars,
amusement parks,
famous beaches, and

almost 59,000 homeless
men, women, children
barely hanging on.

Dream of Peasants' Revolt

So, two guys in suits arrive,
flash their badges,
and say they are from
The Secret Service.

They say the President
wants to fish here
and for security reasons
I must leave.

"But this is America," I say,
"Taxpayers own this river and
I have been fishing this spot
for thirty years."

"We understand,
but you still have to go."
says the short one.
So I go,

but before I do,
I heave five stones in the river.

Salmonspeak

Maybe we never needed gray raceways
and 5 gallon plastic buckets
to propagate our race.

Maybe if we go past concrete hatchery
intake, we will find wild fish
with all their fins

on spawning gravels
used many thousands of years
by the ancient ones.

Sky Whale

After the last whale died
whale-shaped clouds appeared.
Some said "whale spirits."
Others "just clouds."

The ones who spoke the latter
were the same who
killed most of the planet.

Whale Eggs

Today we look past breakers
and apple gray crab pot
buoys are whale eggs.

"Whale Eggs?" someone asks.
Binoculars rise
and a swan lady in white jacket,

yellow sandals, wonders
"So why do they have numbers?"
"Because biologists track the migration."

Her husband, a farmer-type,
says he didn't know whales laid eggs.
"Yeah? Where you from?"

"Wisconsin," they answer.
"Oh," I say.
Their laughter moves

toy-like orange-beaked puffins
from an undercut cliff
to bob for bait fish in the channel.

Charters pass under the bridge,
unload decks
of Chinese-carnival rockfish.

Some Guys

lose a fish
and quit,

spend lives
telling stories

in bars
and parking lots

about the one
that got away,

sleek shape,
blazing eyes

like that's all
they've seen

or ever will.

II. Remembering Rocky Creek

Goats and Feathers

are the kid's favorite things
out here near Mt. Hood
which I guess are better than
most things everywhere else.

Barn

was shelter for goats and horses,
hayloft for lovers,
ark for children.

It was warm in snow,
dry in rain,
until the day it burned

and old men recalled
it was a roller skating rink
built by Crazy Lucy's husband

when Crazy Lucy wouldn't
move here otherwise.

Mushroom Hunting on Black Friday

Crowds of red spawning chinook
and lounging elk
celebrate the day

of autumn abundance,
oak leaf colors,
along a hidden river.

Once, out of habit
I watched a mossy TV
dumped below a cliff.

Far from electricity,
it was home for sow bugs
and centipedes.

It was there I saw my eyes
in silent core
watched by clouds and sky.

Wild Strawberries

near an Oregon creek
were delicious
thimble-sized
unpretentious

shared with many
wild birds who,
unless sick,
wouldn't go near

a city.

14 Sunflowers Across from Blue Scorcher in Astoria

between river and street,
heads bowed,
wait for sun through fog
how many lifetimes?

At Astoria Bistro

On the wall an elk emerges from mist

while many possible worlds sprout
atop a room divider.

A poster Hierarchy of Beards sets off
a hierarchy of sea memories.

Painted wood-carved game heads
stare bemused, in wonder, shamanic.

A spirit-mirror enlarges a festive space
where talk is about art, salmon, lovers,
beloved pets, history of long ago
and in the making.

At New Renaissance Bookstore in Portland, Oregon

A man says I owe him money
from a past life.
Since I am open-minded,
I ask "How much?"
and he says he can't remember.

Mechanic

I call to get my car fixed
and go over make, model, year
but he stops me mid-sentence,
asking "How's your gut?"

"What does that have to do
with anything?" I ask
staring at plum blossoms
and blue breaks through sky.

I recall my Buddhist friend, Ron,
a pipefitter in Prudhoe Bay,
who said "During 1930s Depression
people starved to death

because they didn't know
they could walk outside
and eat dandelions."
I laugh into the phone.

Somehow, I don't know how,
my gut and the car get fixed.

Poem About Edward Hopper's *Night Hawks*

Stories wander in and out like stray dogs looking for homes,
sometimes pairing with human minds, hearts, bellies,
maybe a businessman's war-deformed right hand,
or wounds in memory pond of jilted woman in red
speaking to a man she will lay with tonight who
uses sex as a life ring for a job he hates
as his father, and his father's father.

It will take years before their stories reveal
all of them, in spite of worst fears, are worth loving,
or at least listening to in fully-present fashion,
like hundreds of years ago in the meadow
where Phillies is now, and native men and women,
children on backs,
gathered roots and sang.

Somewhere in Oregon

"There is a story behind
each visible and invisible tattoo,"
he smiles, "but mostly invisible."

"This pheasant was so beautiful
I had to shoot it."

Salal 2019 Event in Rose Center at Lower Columbia College

Before the poetry reading, an opossum emerged from ferns
outside beyond the piano.

How fitting, I thought, *protecting her secret*
and mine.

Poets were humble, authentic, wise
like Ray Carver at Portland State

or better yet, God in a creek meadow
on Oregon Coast.

"I created the 7 universes from scratch," he smiled,
"but almost no one has time to listen."

Climate Crisis

is like the night
at an Oregon campground,
exhausted from fishing,
I returned from restroom
to wrong tent

then woke up
with another family
questioning myself
in dim moonlight
Do I run?

Apologize?
Compliment his wife
or ugly dog?
Maybe they will think
I'm insane, drunk, or worse.

I knew anything I did or said
was woefully inadequate
to scale of error.

Bear Who Nearly Destroyed Earth

"At around midnight on October 25 [,1962], a guard at the Duluth Sector Direction Center saw a figure climbing the security fence. He shot at it, and activated the 'sabotage alarm.' This automatically set off sabotage alarms at all bases in the area. At Volk Field, Wisconsin, the alarm was wrongly wired, and the Klaxon sounded which ordered nuclear armed F-106A interceptors to take off. The pilots knew there would be no practice alert drills while DEFCON 3 was in force, and they believed World War III had started.

"Immediate communication with Duluth showed there was an error. By this time aircraft were starting down the runway. A car raced from command center and successfully signaled the aircraft to stop. The original intruder was a bear."

—nuclearfiles.org

I smelled cinnamon-baked apples in your garbage
and climbed a fence.

I had nothing against bears in Russia,
Alaska, Japan, or China.

I wasn't a republican bear, democrat bear,
Christian, Jew, Hindu, Muslim, or atheist bear.

Our cubs were never taught to hate,
or *Duck and Cover* under logs.

We didn't have war posters on evergreens
or bear radio alerts on rivers or creeks.

Mr. Moose, Mrs. Bobcat, and I were fine
before your kind moved in

killing most of us, like you did Indians,
without learning our ways or language.

Sure, we had occasional floods, storms,
brush fires, lightning, tornadoes, but

we could always count on blossoms
following snow, summer following spring.

We moved according to seasons,
and searched for family, berries, mates, sleep

in mountains, meadows, canyons,
wildflowers and waterfalls

beyond your fearful thoughts.
So, you shot me and I'm dead.

And you almost were.

Future History Book

Survival in short-term Europe
meant cities.

Survival in long-term Europe
meant none.

Tsunami

"Cultures living near the Pacific Ocean in areas hit frequently by earthquakes developed over time an early warning system for a tsunami. Stories and legends are used to warn future generations of the possible risk and, when possible, offer advice how to survive such a disaster."

—David Bressan, "Ancient Stories Preserve The Memory Of Tsunami In The Pacific Ocean," forbes.com, Mar 23, 2018

In days past
when saltwater rushed out
you imagined tsunami
and warned neighbors,
ran for hills

but now
with a big one poised
to wipe out most
of life on Earth
and nowhere to run

to fully escape
many do nothing
while some try to protect
what can be
while there is still time.

In space we are
a blue jewel
in a baby elephant's ear,
waterdrop with reflections
around glowing coal,

eye of nearly
belly-up dolphin
in dark river
splashing to right herself
by almost swimming.

Written After Newly Discovered Fragment from *Gilgamesh* Warned "we have reduced the forest [to] a wasteland [in 2150 BCE]"

"The previously available text made it clear that [Gilgamesh] and Enkidu knew, even before they killed Humbaba, that what they were doing would anger the cosmic forces that governed the world, chiefly the god Enlil. Their reaction after the event is now tinged with a hint of guilty conscience, when Enkidu remarks ruefully that ... 'we have reduced the forest [to] a wasteland.'"

—Marissa Fessenden,
smithsonian.com, October 7, 2015

We must grieve losing the real Eden near Tigris and
Euphrates Rivers
unwilling to stop killing 461,000 Iraqis from 2003 to 2011,
and more widely-reported 4,497 U.S. Military deaths.

Now, our blue jewel planet undoes 4.54 billion years of
evolution
to try again. In a twist on Plato,
there can never be hope until scientists become politicians
or politicians become scientists.

What shall we do as the end of our Anthropocene is
replaced with Roachocene?
Shall we write poems, read poems, eat poems frosted on
birthday cakes, or belly of a lover?

Essex Was and Is

"*Essex*, American whaling ship that was rammed by a sperm whale on November 20, 1820, and later sank. Although all 20 crewmen initially survived, only 8 were rescued following an arduous journey that devolved into cannibalism. The sinking inspired the climactic scene in Herman Melville's *Moby Dick* (1851)."

—Britannica.com

With politicians and merchants involved
it was hard for whaling captains to confess
their 87-foot ship was sunk by a whale.

Nearly 200 years later, it's similarly hard
for billionaires to admit *Our asses*
are being kicked by climate crisis.

On the Alaska Ferry *Malaspina* South of Juneau, July 18, 2019

"No Service"
cell phones remark

and people look up
at the sea

and at each other
as it will be

soon with
the climate

like that song
by Talking Heads

that says "My God!
What have I done?"

Bureaucratic Climate Dream

When I wake-up we are crashing
into snowy peaks
and pilot and copilot are drugged.

I know how to fly
but attendant says
I need official paperwork

and there is no way to get it
at this altitude
so I should just relax.

"My official paperwork
is I know how to fly," I offer.
"No good," she scolds.

The speed of descent increases—
I see a goat chewing a *New York Times*—
as she adds "These rules

are for passenger safety
so if you try to enter the cockpit
you will be arrested, or worse."

She smiles like a TV host
and gently whispers "Imminent death
of your family is not personal."

I try to warn the obvious,
she is going down
with the rest of us, but

she hands me a bag of nuts
then is busy with other passengers.

On a Back Porch in Clatskanie, Oregon Listening to Two People Argue About the Holy Day Being Saturday or Sunday

He says Saturday.

She says Sunday.

They both cite historical
and biblical records

and want me
to take a side.

The argument reminds me
each day is holy,

each place,

insect

larger to us

than us to cosmos.

I Dreamed a Tavern Called The Spirit Fish

where a man by a fireplace told a story
of getting skunked on an Alaskan fishing trip
then returned to find his 9-year-old son
caught a salmon in the creek
behind their Washington home.

"Sometimes our seeking must be closer
to people and places we love."

Remembering Rocky Creek

I

Just north of Miroco's pastel cottages
and inland of the Otter Crest Loop
I follow a worn trail
away from an old logging road
into a hidden gorge
where late afternoon cutthroat trout
slap emerging caddis flies.

Crystal current meanders through fallen logs,
past sun-lit riffles,
fanning across a brown sand delta
of honeysuckle and cattail.
Gone now are Native canoes,
raised totems, cedar huts,
ritual shell middens and potlatches.
Instead, the creek slips through
a graveyard of giant evergreens
and enters an ochre gate
of rusty railroad track
to a concrete tunnel beneath Highway 101.

Above, the man-made wall
casts a shadow on the once-magic waters
like a monument to wild salmon
floating belly-up in crisp Pacific moonlight
many years ago
beneath the spouting concrete culvert
—their ancient way home blocked—
vanishing one by one
like agates dropped into the sea.

I fish the pools in the secluded canyon,
clouds part, and rays of sunlight
strike the crystal waters
like rays in Bible pictures
reminding me of days offshore
trolling for silvers along tide rips
where the blocked creek mouth
first caught my attention.

II

I close my eyes and remember
my days on the *Starfisher*.
As the bow slices water
a wide pink veil
lifts from the rocky coastline.

Wooden boats approach
the frantic diving gulls,
fast and furious salmon.
The cry of "Fish on!"
joins the rush
of the summer morning.

Pelicans cruise in formation
just above the surface.
Puffins bob for bait fish.
Albatross soar on the horizon.

Gray Whales migrate north in pairs,
their 12-foot-wide, twin-fluked tails
rise from the dark waters.
Forty-foot breaching loners
vault again and again above the surface
like huge hooked rainbow trout.

Orcas chase schools of silvers.

Dorsal fins of blue sharks
slice V-paths in lake-calm waters.
Flat silvery ocean sunfish,
as large as an engine hatch,
bask like strange creatures from outer space.
Tropical sea turtles lazily paddle
in warm offshore current.

III

Back to the present,
I rest on a mossy boulder,
clip the caddis fly
and tie on a stone fly nymph.

A few remaining
Sitka spruce trees, 150-feet or so tall,
stretch up as if to see over the wall.
Evening wind flows
down from Cape Foulweather
and through the Sitka Spruce,
leaves its tracks on pools
shining in the canyon.

Quietly, it asks me,
"Who erects this barrier
where there should be a bridge?"

In the Pacific Ocean, land and sky
migrations of men
replace the ancient ones of animals.

New condominiums—
Sea Ridge, Thundering Shores,
Little Whale Cove—
change the face of the coast,
crowd pastel cottages
and open spaces along the sea.

Gray concrete pools
in modern salmon ranches
at Yaquina and Coos Bay
substitute for cascading creeks
such as this one, where hook-nosed bucks
and plump ripe hens
once planted the secret life of the race
beneath autumn leaves and pebbles
to burst forth again with spring rains.

It is nearly dark. I stand knee-deep,
roll-cast the nymph
along a bank of submerged logs.
There is a vibration in the rod tip.
I set the hook, ratchet the reel,
guide the half-foot fish to shore—
crimson throat slashes,
bright orange black-spotted fins,
olive back and rosy gill plates.
I remove the hook from its lip
and it darts away,
slips quickly
into cover of gnarled roots.

Fishing at Age 50

Wading swift gin-clear crystal
beside deer tracks,

with a rusted Quick reel
still usable if oiled,

under spring's dragon clouds
like when I was 5,

my last salmon is
washed and gutted

with his heart beating
on mossy stone.

Before Shura died
she donated her gray-blue

artist's eyes
to help a blind person.

By the rose garden
in Ladd's Addition

she said, "After I'm gone,
where the real is—

I'll meet you there."

Thinking of Climate Refugees on a Remote River in Oregon

Wind gust pushed honey bee in river,
too far out for me to help.

I stopped and focused on struggle
as current took her down rapids.

From her frantic legs, twisting body and head,
it was clear she wanted to live.

I wonder how many North Americans
eating honey think of this.

Sand Meditation

One morning I wrote her name
in wet sand at low tide
and by noon it was gone.

Next, I wrote my country
and by evening
it was gone.

Soon, even sand and tide
were gone
as I was

circling back like an extinct
whale, salmon, goose
with nothing to be

and nowhere to rest.

III. In Praise of Wild Rivers

Above the Hatchery

wild steelhead spawn as they have 10,000 years
and elk follow ancient trails through sword ferns.

Rodless, rifleless, I watch, knowing what humans
have become downriver, and could be again

with song-dance-mask-story ritual,
gatherings in cedar longhouses

and dreams connecting us to everything up here
and beyond.

Written in the Dark Near a River

Ancients knew
rocks have hands,
rivers have eyes,
trees breathe.

It is no mistake
I feel at home.

My People

love to fish, hunt,
tell stories about
ones that got away.

Know names of rivers,
rock formations, and trees
in our territory.

Listen to outsiders
with caution
and generosity.

Pass down
what we know
about local legends.

Most are born here
and stay
or return

with tales of
faraway men,
some good,

but others that try
in their madness
to kill you

like they drank
from a poisoned well.

At Hermosa View Elementary, 1968

I did not like my first day of school
with Ms. Letnich
coloring red, yellow, green circles

on a fake shoebox traffic light,
taught reality was more in human
symbols than nearby smiling dolphins,

blazing bonitos, or long-haired
young ones with guitars above tides
and shifting sands,

forever seeking
lands beyond
what we see, and what there is.

Memory

Once, during 7th grade football practice
my team had to help corner runaway horses.
Nostrils flared, eyes shot lightning,
but there were five of them and forty of us
yelling, stomping, trapping
until owner arrived with trailer.

That night I dreamed 10,000 years back
when St. Anthony's school was a stand of red cedar
and the church a meadow.
Horses were friends.
We learned about healing herbs,
prophecies of masking and unmasking.

Oregon City Flood Year 1964

rivers took back streets
and patterns of our lives
were set by clouds.

Tony's Fish Market
sold from a rowboat
and neighbors sandbagged homes.

Salmon crisscrossed highways,
easy picking for a pitchfork
or kids on hands and knees.

Maynard's Love

meant life jackets for his milk cow
during the flood
and howling like tire chains in the wind
when she died.

He sported neon lime coveralls
and one
slightly irregular eye
so farm kids made fun of him

but his capacity to care
awed the rest of us into believing,
despite everything that happened,
being human was still possible.

In This Country

After 30 years
of hard work
and sacrifice,
meaningless ego,
I achieved
the high honor
of being a gardener
and dog's pillow.

Man and Black Lab by Fire

10,000 years hunting together
on every continent

and keeping company through
storms, hunger, loss.

Fishing buddy in
all kinds of weather.

They survive in a world
that takes and takes

and rarely gives.
This is the one photo

space aliens will keep
on their walls

recalling Earth that was,
and could have been.

Wapinitia Poem

"The November's full Moon was called the Beaver Moon by both the colonists and the Algonquin tribes because this was the time to set beaver traps before the swamps froze, to ensure a supply of warm winter furs."

—*The Old Farmer's Almanac*

"Wapinitia — This is a Warm Springs word [. . . that] suggests a location near the edge of something."

—*The Bend Bulletin*, Aug. 20, 1952

One day my wife calls
to say she's giving up
London fog drinks
to save money for our old age.

The next, Les Schwab says
she needs $2000
for new tires and brakes.
I tell her "I understand
London Fogs,
but I'm not sure about Les."

Later, an unexpected branch
crashes through her windshield
on McLoughlin Blvd
calling me home across
Wapinitia Pass.

We'd like to think
we control destiny
but Beaver Moon
says otherwise.

Lights in Doug Fir

may be stars,
UFOs,
cougar eyes.

All night
I rest
on river stones

on a wildflower
and fir trail
ledge

to get my
salmon spot
in morning.

Whatever
the lights are,
I celebrate

being alive.

Poem for Slim

"If the fish is in the river,
he's in my bucket," says Slim.

I wish I had that confidence
about anything,

changing jobs,
planting garlic,

letting go
what I must let go.

Cassie

was infamous for shouting
"You don't love that woman!"
at a park wedding, and being right.
Another time she said
"That lab's gonna run away
if you don't walk it," and it did.

Once, she stopped me
by the store's frozen foods
to say hell is the absence of love, and
the whole Columbia Gorge will be underwater
if we don't stop sinning.

Another morning, as I was going fishing
she surprised me as a voice in the dark
hissing behind a fir:
"How can you know a river
if you don't know yourssssself?"

The Wolf in Estacada's Safari Club

After fishing the upper Clackamas River in 1979
my high school friends and I ate chef salads
among stuffed exotics shot by Glen Park in six countries.

Dozens of lions, leopards, tigers, bears, monkeys spoke
in our imaginations and dreams
of where we'd go in the world.

Of course, early deaths, divorces, and varied jobs
separated us farther than characters in *The Big Chill*
but even now, in our 50s, a few of us return to the river

for summer and winter steelhead,
fond memories of times more innocent and hopeful
before flesh met steel and lost.

My wife's sister eventually bought the place
and left the animals with her Wyoming cowboy father.
Up the road, canyons below Mt. Hood are painful, but
    rewarding.

Neighborhood kids, not much different than us
in our prime, knock asking if they can come in
and pet the wolf.

Names

In Oregon
Mill City has a mill
and the Salmon River
has salmon

which may explain why
people are happier
than other places
I've lived

and fished,
where names of places
are just names
connected to nothing

and city folks drive to work
in new cars
shiny as coffins.

Hat Haiku

That can’t be my hat
on the stone by the river.
There’s gray hair in it.

World Politics

Family at next table is upset
they didn't get enough cherries
on their pizza
so they want a refund but
waiter refuses.

The alpha is a man of means:
"See this gold watch?"
Waiter: "Whatever, Dude."
Any moment I imagine a SWAT team
getting this man his cherries.

Half a world away in India
it's 122 degrees Fahrenheit
where people add mist to bed sheets
to cool off,
and wear wet towels.

Lacking electricity,
they pay the most for warming climate,
cooking with dried cow dung,
living and working in heat
that would keep North Americans inside.

Life, like art, is always about perspective,
one family at war over cherries,
another, far from headline news,
just praying for rain.

December Sun

Most of the world isn't waiting for your poem.
Instead, it wants you in a service job
as an underpaid nurse or teacher,
farm worker, child care professional,

paramedic, home health aide,
social worker, or taking orders from
harried people in Wendy's, Burger King,
McDonald's, Taco Bell,

obeying all laws, paying taxes,
walking when the light says walk,
stopping when it says stop,
leaving birth and death to so-called doctors

and insurance companies
compassionate as thrown stones.
Write anyway.

In Praise of Wild Rivers

Computers write poems,
and scientists make artificial hearts.

Robot dogs sit, fetch,
and open doors.

Soon Amazon Prime will deliver
to Moon and Mars

but will be unable to find
my home in the Cascades.

On a stone beyond cell range,
I think of my ancestors and reflect

as long as wild rivers remain
I remain.

Turkey Feather

Maybe death is like that day on Hidden River
a wind gust caught my hat
and I was so into catching steelhead
I almost didn't notice a green felt Stetson
that looked just like mine
snaking around boulders through rapids.

Lovemaking After Fight With My Wife

Like two salmon battered
and scarred
arriving in fern grotto

pulled by spark
to mate and rest hovering
over ancient gravel.

Camping on Lower River

Five salmon in my cooler
passed by here
intent on spawning.

I reeled them in,
intent on eating
and living.

Sex and hunger
meant hundred-yard dashes
underwater,

screaming gears
and a priest-club
to give last rites.

Some big ones
got away
to keep the story moving.

Net Pen Tuna

"'They are in that round pen for so long that all they know how to do is make left-hand turns,' [. . .] says [an "anonymous" worker]. PVC-and-net pens broke open during the recent storm, freeing hundreds and perhaps thousands of bluefin. 'The big surf drove them in to shore.' [. . . .] Benson says there were other fishermen on the beach that day hand catching the bluefin as they washed up in the surf."

—Dave Good, "Bluefin Tuna Wash Up On Imperial Beach,"
sandiego.com, January 27, 2010

Wild bluefins are most prized,
deliciously wary, fleeing boats' approach,
but net pen cousins are dumb as bilgewater.

I'm writing this at Village Grind in Wrightwood, CA,
under snowy peaks, imagining who humans were
before we were put in K-12 boxes.

I think of Ishi's resilience and grace.
"Purple mountain majesties."
Undiscovered rivers and creeks.

When I was 15 I searched
for hidden treasure
in Wyoming's Vedauwoo rocks,

green-speckled
fire-side brook trout
in nearby beaver ponds.

I dreamed sea monsters sketched
by men so long offshore
they confused dreams and reality.

290,000 years we were wild
until Enkidu traded his animal friends
for cities and language.

Language, he thought,
harlot in his arms, *will bring us*
close to each other, and sometimes it did

but mostly it was a tool for trickery,
lies, distortion,
and exploitation of poor.

Here, rise faint guitar music
and soft-scented pinesmoke.
I think of Slim Bracken, 79, my logger fishing bud,

who drove my wife Suz
"on a quiet ride through hills"
100 miles an hour

and walked cliff edges where one slip
meant instant death.
He died a year ago crashing his bike.

I'm close to retirement
telling writing students
"Just for today, forget about thesis,

grammar, organization, purpose, audience.
I'm tired of all that.
Give me something, anything, wild."

Dream of Customer Service

The woman behind the counter asks
if I found everything I wanted.

"Actually no," I say. "What I
wanted was love."

"Officially," she replies, "we're
out of stock."

"But unofficially," she whispers,
"everyone keeps a little in the back."

"Pigeons Can't Fly"

says Professor Ted,
"It's an optical illusion."

"Tricksters," I say,
having learned

never argue
with Seattle's homeless,

reaching past emptiness
of gray sea and sky

for the one gift I have.

Eulogy for Jenny Copper of Hood River, September 10, 1929—March 25, 2013

The first time I met Jenny in her 70s
she asked me in my 40s
if I was married.

When I said no she added,
"There are only two women
in this county fit for marriage,
and I'm one of them.
Are you interested?"

I said I'd have to think about it.

"That's just your answer for everything,
isn't it?" she snapped.

I had known her less than three minutes
so I laughed, and she laughed harder.

I gave her a steelhead I caught
below her orchard
and she gave me three
vine-ripened cantaloupes
I never forgot.

Lingcod

Wolf eel on steroids.

Nightmare demonfish.

Dragon-toothed beast

delicious when baked

at 350° with lemon butter

like any fear

caught and filleted.

A Spell of Birds and Fish

Tonight, I hear the song
of an unseen bird
who remembers the way
across the canyon,

carrying life
that need not see itself
to know its beauty.

I see a bristlecone sunset
like the side
of a huge fish
passing by,

early stars
the eyes of those
who broke free.

Drum Circle

My name is not on granite or bronze
and doesn't have to be.

Long ago, it was carved
in the origin of the Columbia Gorge
and universe

same as any thorn,
honey bee, or sword fern.

Here

secret script waits
through flowers and skulls,
song and grief,
ice ages and climate crisis,

but most men and women,
too busy to read it,
live as shelled mollusks
waiting for hands, teeth,

incessant beaks of gulls
to pick at their remains.
What they don't know is
in galaxies before

were the same words
and challenges
rooted in place, not person.
Start with facts—

in a billion years the sun
will evaporate Earth's seas,
and in 4 billion a bigger galaxy,
like a bigger fish,

will eat ours.
While there is time
don't squander
what has been gifted to you.

Wild Trout

don't care about your name,
friends, or income.

The car you drive and college
are irrelevant.

Instead, trout want to know
how much water you've read,

how often you listened,
what you saw and

every thought you had
in your time alone

on the hidden river.

Photo by Suzette Starbuck

About the Author

Scott T. Starbuck's book of climate poems *Hawk on Wire* was a July 2017 "Editor's Pick" at Newpages.com and was selected from over 1,500 books as a 2018 Montaigne Medal Finalist at Eric Hoffer Awards for "the most thought-provoking books." His book *My Bridge at the End of the World, New and Selected Poems* was a 2020 Finalist for the Blue Light Press Book Award. He taught ecopoetry workshops the past two years at Scripps Institution of Oceanography in UC San Diego Masters of Advanced Studies Program in Climate Science and Policy. Starbuck had residencies at Sitka Center for Art and Ecology on Cascade Head, Artsmith on Orcas Island, as a Friends of William Stafford Scholar at the "Speak Truth to Power" Fellowship of Reconciliation Seabeck Conference, and at PLAYA near Summer Lake, Oregon. His *Trees, Fish, and Dreams Climateblog* at riverseek.blogspot.com has readers in 110 countries.

A longtime Pacific Northwest resident, he lives in Battle Ground, Washington.

CPSIA information can be obtained
at www.ICGtesting.com
Printed in the USA
FSHW011223210421
80548FS

9 781936 657582